Introduction.

Having lived with depression since I was about fifteen years old, I have been incredibly lucky. Lucky may sound like an odd word to use with regards to depression, but I have been through and dealt with many aspects and areas of depression and I am still here. This in turn has taught me countless valuable lessons about life and mental health.

The very first lesson I learned about depression is that most people will do anything they can to keep it quiet. Whether they have it or someone they know, they don't want to talk about it. We have all been guilty at some point of thinking that if we ignore a problem it will go away, but it never does. If we ignore a problem it gets worse and worse until we can't cope any more.

Mental health issues are hushed up as they always have been. People don't want to talk about them because society seems to say "if you have a mental health problem then you are crazy" but you're not.

And whilst talking about mental health issues may not cure them, it will make those that live with them feel better. Feel better about speaking out and seeking help and feel better knowing that they are not alone.

In this book I aim to cover some of the aspects of depression and the ways they can be dealt with and to encourage people to talk about depression. There's nothing wrong with having depression, it can happen to anyone and the very fact that people don't want to talk about it could mean that someone you know is going through it right now.

I'm not an "expert" on depression and I don't claim to know everything about it. Everyone's experience of depression is different. I have just lived with it for a long while and I want to help anyone else that is struggling if I can.

Be honest with yourself.

Coping with depression can be an incredibly frightening time, you may be experiencing thoughts and feelings that are completely new to you and the chances are you will feel alone and frightened. But when it comes to coping with clinical depression one of the most important things to remember is that you are not alone. There are people that you can talk to and people that care about you, no matter how alone or lonely you might feel, someone cares.

But before you can even think about admitting it to anyone else, you need to admit it to yourself. You need to acknowledge that you have a problem with the way you are feeling and that you can not cope. I know how hard this may sound but if you do not do this then you can not move forward. If you do not truly admit how you feel to yourself then you will be living in denial and nobody will ever be able to help you.

But before you admit it just let me clear up a few things.

1. Just because you are depressed does not mean you are crazy.

2. It does not mean you are weak or that you have a weak mind.

3. It does not mean that you will have to be on medication or in therapy for the rest of your life.

    4. It happens to a hell of a lot more people than you think.

5. You are not alone.

Why don't people want to talk about depression?

Depression is one of, if not THE most commonly treated illness in the world . And despite the fact that it is one of the most commonly treated illnesses in the world and that just about everyone will be affected by it in some way, nobody talks about it! Why?

We would tell somebody if you had a cold, or a chest infection or mostly any other kind of illness but why are people so afraid to talk about their mental health?

There are various reasons why people do not want to talk about their depression.

⅄   Because society tends to label anybody with mental health problems as "crazy"

⅄   They are afraid people will think they are mentally weaker or stupid

⅄   They believe they will be on pills for the rest of their lives

⅄   They are afraid to talk to anyone about it because talking about depression is not the "done thing" and so they sweep it under the rug, very often getting worse and worse.

⅄   They feel that people will judge them for not being able to cope with life.

⅄   They think. they will have to have therapy and many people still believe therapy is for "crazy" people.

⅄   They feel like they have no one to talk to because they think no one understand what they are going through.

Whatever the reasons, the fact is that people still sweep mental health problems under the rug and don't talk about them, when ironically talking can be one of the most helpful things. Feeling like an outcast is one of the most isolating things about depression. It leaves you feeling alone and that nobody understands what you are going through.

People are made to feel they are not "normal" which is really unfair since having depression is statistically very normal!

I realise that attitudes will not change overnight, but they do need to change. People who are living with depression don't need to be made to feel any worse than they already do. They don't need to feel alone and isolated. They need to feel accepted and like they can turn to people for help, advice or just an ear.
So please, if you know someone that is living with depression, ask them how they are, they may not tell you, in fact they will probably just tell you that they are "fine" but at least they will know someone cares.

What depression isn't.

Can you tell me what depression is? Google it and you will find two very different definitions

1.Severe despondency and dejection.
2.A condition of mental disturbance

Clear as mud.
And the reason there is no clear definition is that depression is not a clear illness. Being depressed means different things to different people so it can be hard to put that into words.
But depression is NOT:

*Something people choose.

*Something that people can just "get over" by being positive

*Weakness or weak mindedness

*Selfishness or self pity

*Insanity or craziness

*Something to be ashamed of

These are the kinds of myths and narrow minded prejudices that cause people to feel ashamed of their situation and not seek help when they so desperately need it.

Just because a person is depressed doesn't mean that they are any less of a person than you are. It does not mean that you are stronger or smarter than them.
Only an idiot would believe that someone could live their life with depression and still be weak. It has made me a stronger person than I have ever been in my entire life.

The "snap out of it" mentality seems to be shared by a great many people and with the greatest of respect is possibly one of the most stupid things to say to a person suffering with depression. If they could snap out of it they damn well would! I don't care what anybody says nobody chooses to be depressed. It can not be snapped out of, it is something that takes time to deal with.

Depression is not insanity, insanity is defined as doing the same thing over and over and expecting different results. So maybe by that definition, depressed or not, we are all insane?

What depression can do.

So many people still believe that depression is just something that makes you feel "a bit sad" or "down" and that it can be cured by being positive or by sheer force of will. But don't you think that if people could get rid of it and be positive then they would be?
Depression can:

- Leave you feeling helpless

- Alone

- Frightened

- In despair

- Make you truly believe that no one understands or cares about you

- Make you think that nothing is important and that life is not worth living

- Take away the passion and bravery you once had

- Turn you into an entirely different person than you once were

- Make you lose sight of everything good in your life

- Make you feel as though your life has no purpose

- Make you feel like an outcast and a freak

- Make you feel weak

- Make you feel as though everyone thinks that you are crazy

- Take away all of your happiness and positivity

- Leave you feeling that people would be better off without you in their lives

- Make you genuinely want to be dead

- Ruin your entire life

- Make you try to kill yourself

These are just a few of the things that depression can do to a person, depression is not about feeling a bit fed up or that you're having a crummy few weeks. Depression is a serious condition and the sooner people realise that and start treating it like one the more people can be spared the soul destroying loneliness and misery that depression can bring.

People take their own lives every single day because they don't seek the help that they need. Whether this is because they think that people will judge them or call them crazy or because they don't know who to turn to this is something that needs to be addressed. It doesn't matter what the reasons are it needs to stop. The stigma needs to be removed from depression, and people need to understand that just because you can't see it doesn't mean its no very very real.

How it makes others feel.

I have talked a lot about how depression takes those that have it, but they are by no means the only people that are affected.
Friends and families can be torn apart by one person's depression, relationships have been shattered beyond all repair and family bonds totally severed, all through a lack of understanding.
It can make others feel

*Frustrated (they don't know what to do for the best and don't know why you don't just snap out of it)

*Frightened (they are scared for your safety/sanity and feel like you won't be the same person any more. People often say they feel like they are losing someone to depression)

*Helpless

*Alone

*Hurt

Depression is the same as any other illness, it can affect everyone around the person that has it.

The symptoms:

As I have previously explained, there is no clear definition of depression which means there is not a definitive list of symptoms to look for. But there are a few warning signs that could possibly be depression.
In no particular order, these are:

*Tiredness

*Irritability

*Low moods

*Mood swings

*Lack of interest in doing things

*Lack of personal care (hygiene, nutrition etc.)

*Distant (the person may seem miles away)

*Avoiding (they may avoid people they previously liked)

*Dishonesty (they may lie to avoid talking about what's bothering them or to avoid an unpleasant situation)

As I said these are just a few symptoms that could point to depression, there may be many others but if someone you know is experiencing more than three of these it may be worth seeking some sort of medical advice.

Telling someone.

Telling someone that you are suffering with depression isn't easy, the judgemental stigma that is attached coupled with peoples misunderstandings can make talking about it seem like the most terrifying thing in the world.
I understand that depression can bring with it a certain lack of trust, you feel as though no one will understand and everyone is going to judge you. But they're not, if you feel there is no one from your friends and family you can trust enough to tell then there are other people to talk to.

Your doctor or nurse practitioner are there to help, that is their job. They will not judge you or think that you are crazy, you won't be sectioned or laughed at. They will not tell anyone else, they are not allowed.
If you feel talking to your family doctor will be too hard, you can ask to see someone else, the surgery is obliged to abide by your wishes. They can not force you to talk to someone you don't want to.

Alternatively you could talk with your nurse practitioner, the nurse practitioner at my local surgery was a total godsend and I don't think I would be here right now without her.
There are also numerous helplines and chat-rooms bursting with people that are going through the same thing you are. Don't underestimate the importance of other people's experiences, they have been through what you have and they have lived to tell the tale, let them help.

Helplines are a brilliant source of help, you can talk to someone any time and one of the best things about them is you can't see them and they can't see you. This is not for everyone but some people find it easier to talk to someone that can't see them.

They are there to listen to you in total confidence, they will never judge or tell you what to do. They may advise you of the best course of action but it is totally up to you.

When telling anyone the most important thing is that you are totally honest, whether it is that you are not sure what is going through your mind right now, or that you have had thoughts about taking your life, be honest. The only way anyone can help is if they know what is going on.

Help.

There are two main kinds of help for depression, medication and counselling or therapy. And one of the biggest myths about the medication aspect is that once you start taking anti depressants you will have to keep taking them for the rest of your life. And while some people do remain on them for a long while, doctors are keen to ensure that you are not kept on them forever.

After a while of you being on your medication, if you seem to be making progress then they will start to consider when you should come off the pills. This is normally done in the spring/summer months because it is a time when people tend to feel happier and are not affected by things such as S.A.D. They will gradually lower the dosage and eventually tell you to stop taking them. You will never be asked to simply stop taking them straight away as this can do more harm than good.

Always make sure your doctor assesses when you should stop taking the pills, this is a decision you should not make on your own. Just because you have been feeling better for a few weeks or even a few months does not mean you can stop taking the pills cold turkey. Any alterations NEED to be handled under the supervision of a doctor. They will reassess your situation on a regular basis and keep track of your improvements, and they will tell you when you are ready to come off the pills.

Counselling or therapy is often used in conjunction with medication. The pills allow you to take full advantage of the counselling.

Whether you are sent to a counsellor or a therapist there are a few things you need to know.

• The doctor has NOT sent you to a professional because they think you are crazy, they have sent you because they know you need someone to talk to that understands what is going on in your mind right now.

• Everything said to your counsellor or therapist is treated with total confidentiality.

• You will only be seeing a therapist temporarily, you won't have to be going for the next ten years.

• You won't be asked to lie down on a big leather couch.

You won't be asked your innermost secrets on your first session, it takes time to build up a relationship with your counsellor or therapist, they are fully aware that a bond has to be formed before you will begin to trust them. This is why most counsellors will give you the option of requesting someone else if you are not comfortable with your present counsellor.

You also have the right to see a counsellor or therapist that is the same gender as you, I saw a male therapist though and he was fantastic so keep an open mind but if you are uncomfortable with seeing an opposite gendered person, tell your doctor and they will arrange someone else for you.

Medication or therapy?

Your doctor will decide on the best course of action for you by trying to establish what is causing you to feel this way. There may be a specific reason or there may be no reason that you can think of. Either way the doctor will use this information to help you.

Your doctor may decide to start your treatment by giving you medication. Many people are wary of taking pills to treat depression because they believe that once they start taking it they will be on it for the rest of their lives. This is not the case, doctors are just as cautious about prescribing medication as you are about taking it. They know the potential risks and will carefully monitor you whilst you are taking the pills and find an appropriate and safe time for you to gradually stop taking them.

When you start to take the pills you may begin to feel impatient because you want the pills to make you feel better immediately, but unfortunately they don't work like that. They can take anywhere up to three weeks to have any significant effect on the way you are feeling. But you must persevere with them, don't give up just because you don't feel back to your old self after a week. These pills need time to work, these feelings didn't turn up overnight and they won't go away overnight either.

After a few weeks you should gradually notice a positive change in your mood, but if after six weeks you still feel no different then talk to your doctor again, they may decide to increase the dosage or they may be able to find another form of medication that will work better for you.

I am not going to lie to you, many of these medications have side effects that can put people off persevering with them but don't let the side effects beat you because they are only temporary and it's definitely worth putting up with them in the long run.

If you find the pills are disagreeing with you badly then consult your doctor again, different pills work for different people, it's a case of trial and error to find out which ones will work best for you.

During this time when the pills have not quite taken effect you may begin to feel vulnerable and scared, this is quite natural but remember there is nothing to be scared of, what you have done is a very brave and incredibly positive thing. And even in your darkest hour you need to remember that despite what you may think at the time, you do want to survive, you do want to live. You sought help and this means that you want to find a way out.

It's well documented that during the first few weeks of taking anti depressants many people will experience dramatic mood swings, feelings of anxiousness and feelings of deep depression. Do not ignore these feelings talk to someone about them, there is always someone to help whether it is a friend, a family member or a help line there is always someone to talk to.

Therapy.

Your doctor may decide that the best course of action is for you to see a counsellor or therapist. Many people are reluctant to take counselling or therapy because they think it's only for "crazy" people. But talking to someone face to face that will not judge you and knows how to help you work out all of your confused feelings can be so helpful.

Unfortunately it is not as simple as your doctor referring you and you getting an appointment within the same week, it doesn't work like that. Since resources are so limited many places have waiting lists of up to two years! But your doctor will know where to send you to get you an appointment as soon as possible.

If you don't know what to expect from therapy or counselling then don't worry, these people are there to help you, many people think that therapists are judging them on every little thing they do, but this is not true, they are simply trying to figure out the best way to help you solve your problems. And don't worry about being made to lie on a leather couch, you can sit down, stand up or walk around the room, it is entirely up to you.

Your counsellor or therapist can not force you to talk about things, if they tried it would be very counter productive but it really is best to try and talk things through with them. You may find it hard at first and any therapist will know that you may not be up to baring your soul in the first few sessions. But you should gradually find yourself building up a relationship of mutual trust with them which will make you feel more comfortable.
As I have said before, having someone to talk to that you know will not judge you can be incredibly helpful for anyone suffering with depression, so if your doctor suggests it, give it a try, you have nothing to lose.

If your doctor has recommended that you see a counsellor and take medication, don't be alarmed, this is a good thing, the pills can help you use the therapy more effectively.

Side effects of the meds.

One of the most common reasons people stop taking anti depressants is because of the side effects that are associated with them.
These can vary from person to person but the most common side effects of anti depressants are:

*Mild headaches

• Sight abdominal discomfort

• Changes in bowel movements

• Slightly more aggressive behaviour than normal

• Temporary increase in sad or negative thoughts

• Fluid retention

• Indigestion and/or heartburn

• Tiredness or dizziness

• Shaky hands

• Sleep problems and vivid dreams.

• Mild memory problems and/or lack of concentration.

As I said before the side effects will vary and you may experience one or two of these but you may experience none. Just because its on the list doesn't mean that you will get it, the list is only there because they are POSSIBLE side effect. But no matter how unpleasant these side effects are you must persevere with the tablets because they will help you so much.

Most people only experience mild side effects but if you find that the effects that you are experiencing are severe then you must consult your doctor. Do not simply stop taking them because this could prove dangerous. Your doctor will be able to find you some medication that suits you much better, so don't think that if they have prescribed you one certain tablet then you have to have it no matter what. Your doctor will talk to you and find you the right medicine. Remember that these side effects are only temporary and they will gradually get better as your body becomes used to the effects of the medicine.

.

How people treat you when you're depressed.

It seems that as soon as someone finds out you are suffering with depression, they immediately start to treat you differently, Walking on eggshells every time they see you, never daring to ask the dreaded how are you question for fear they will be given a gruelling account of your mental state.
And to be fair most people can't help it, because it is never openly discussed people are afraid to ask the wrong questions.

And it's not that people don't care, they simply don't UNDERSTAND and how could they if nobody talks about it?
And this can be an incredibly frustrating thing if you are on the receiving end of the "kid glove approach" It can make you feel as though they are treating you like an unstable person who may snap at any moment and turn into a violent psycho!

And as annoying as this is, it's not their fault they just don't understand how to treat someone who is depressed.
And for those of you not suffering with it I will tell you how you treat someone who is depressed........ Exactly the same way you treat everyone else!

Up and down days.

We all have them, depressed or not, but if you do happen to have depression then they are that much worse. But that does not mean that if you or someone you know is having a bad day the they are immediately going to have a downward spiral. I'm not saying it doesn't happen because it does, but sometimes people are just having a bit of a slump. It's perfectly normal and happens to everyone dealing with depression.

And even if someone is a bit down, or having a bit of a slump it doesn't necessarily mean that it has anything to do with their depression. It could be that someone or something has really annoyed or upset them, someone I know always blames my depression if I get angry or upset with anything or anyone and never fails to ask the "have you been taking your pills?" question and it makes me want to scream! Life makes you upset and angry sometimes, not everything is about the depression.

Dealing with down days is something that will be different for everybody. But once you find something that works, that lifts your spirits and makes you genuinely smile then keep hold of it. Don't ever let anyone tell you it is stupid or a waste of time, people make fun of me for liking Doctor Who as much as I do, but I know how much more positive and upbeat I feel by the end of the episode.

And if someone you know with depression develops an interest in a certain thing, and it clearly makes them happy then leave them be and let them be happy!

Relapses.

It happens to just about everyone that suffers with depression, you are feeling as though you are on the right track, only for the feelings of negativity and despair to creep back up on you.
You may suffer a mild relapse or you may find that it is more severe, but the most important thing to remember is that it is temporary.

You may feel like you are getting nowhere and that carrying on with treatment would be pointless because it isn't helping, but it is. The treatment and help you have available to you will be the thing that gets you through this, the thing that stops you from giving up altogether, or worse.
Not everyone will suffer a relapse but enough people do to make it an important issue.
There is no doubt about it, relapses are hard. And the reason I think that they are so hard is because you genuinely feel like you are getting better for a while and then that feeling is snatched away leaving you feeling alone and scared again.

If you begin to feel like this then it is certainly worth talking to your doctor or nurse, they will be able to advise you of the best thing to do in your situation. Without the nurse practitioner at my local surgery, I really don't think I would be here now.

And the same as before, if you feel like you can not talk top anyone then simply pick up the phone and call one of the many help lines available, they are there to listen to you.

Only you can recognise the signs of slipping backwards because they will be individual to you, but it is important to take action as soon as possible, the longer you leave it the further back you could potentially slide. Nipping it in the bud will be easier than leaving it, I promise.

Staying positive.

Living with depression can be one of the most trying times of your life. It has a profound effect on not only the way you behave but also the way you think. And if you are unfortunate enough to suffer with depression then negativity can very easily take over your entire life. Sometimes you may feel as though there is no point in doing anything at all because it is simply not worth it.

Positivity can be an incredibly powerful tool when dealing with depression but it is not easy to know how to be positive when you feel so bleak.

One of the most important things to remember when trying to remain positive when suffering with depression is that you are not alone; there is always someone you can talk to. Whether it is a family member or a complete stranger on the end of a telephone, there will always be somebody that will listen to you. Take advantage of this and talk to somebody.

According to a study by the NHS, talking to somebody significantly improved the general feeling of positivity and self-worth in patients suffering with depression.

Depression is an incredibly upsetting thing to live with and so if you feel as though you need to do something to cheer yourself up then do it. No matter how bizarre it may seem. Whether it is dancing around the room to your favourite song or having a cup of coffee in your favourite café, if it makes you feel more positive about your life then do it.

A study conducted by clinicaltrials.gov showed that physical activity had a significant effect on patient's levels of depression. Thirty minutes of moderate exercise per day was shown to leave patients feeling more positive and happier.

Make a plan of how you can improve your life and the way you feel about it. Write everything down that you think would help you to improve the way you feel and write down what you want your life to be like. Make structured bullet points and be clear and concise. This will help you to feel that there is more structure in your life. It is also something that you can refer back to in order to see how you are progressing. Many people turn to self-help books as a way of feeling more positive when they are depressed. Browse your local book store or library to see if something catches your eye.

One of the best ways to help you feel more positive when you are depressed is to remind yourself that you will not feel like this forever, it will get better.

Emotions.

Keeping control of your emotions is difficult at the best of times but when you are living with depression it can sometimes be almost impossible. This is possibly because depression brings with it a certain lack of perspective.

A therapist taught me a really great way to help keep control of your emotions and the brilliant thing about it is that it works almost immediately. Whenever you feel yourself losing control of your emotions or losing perspective then try looking at your situation differently.
Take a step back and think, how would someone else looking at the situation react?

Would they be as hard on you as you are being on yourself?

Would they be upset by what is upsetting you?

It is a really useful tool to help you regain some control and perspective.
Another great way to help manage your emotions is to make notes of any times you feel low, angry or upset. Write down the time and what caused you to feel this way (if there is a reason) This will help you to identify any patterns that could potentially help you to avoid stressful or upsetting situations in the future.

Try to picture your feelings as an object (it can be any object you like), then imagine yourself picking that object up and placing it in a very big, completely empty box. Then when you feel calmer and more prepared, take the object out of the box and deal with it in your own time and in your own way.
These are just a few of the ways that can help you to regain some control over your emotions, your doctor or therapist will be able to offer many more.

Keeping track.

Keeping track of how you feel when you are depressed does not sound like much fun, but it can really help. It can help you keep a track of your progress and it can also help you see if there are any particular patterns to your moods (sometimes there are, sometimes there are not)

It can also help your doctor. It can help them track your progress, and it can also help them identify any patterns that you might not be able to spot.

You can be as specific or as detailed as you like, but a system that has worked well for me is to simply use the numbers 1-10, 1 being the lowest mood, 10 being the highest/happiest.

If keeping feelings diaries or journals is a bit much for you (and each to their own, journals are not for everyone) this can be a quick and easy way of keeping track of how you feel.

Divide each day up into morning, afternoon and evening and simply write your number in the right time slot. And I'm well aware that moods can change very quickly so you can put more than one number in each slot.

If there is a reason for a sudden number change then it's best to make a note of it, just for posterity.

It might seem useless at first, but looking back over my "chart" at the end of a week really gives me a sense of organisation.

How to help a family member with depression.

There are many myths that surround depression; the biggest being that if you are depressed then you must have a terrible life. This is not true, in fact some people that do suffer with depression have quite comfortable lives and are often surrounded by people that love and care about them. Depression can strike anyone at any time. But when a family member is suffering with depression it can be incredibly hard to know how to help them. But if you want to help them through it then you need to know how to be there to support them.

The most important thing you need to remember if you want to help your family member coping with depression is not to judge them. They cannot help feeling like this and you need to remember that when you are helping them to cope. Judging them will only make them feel worse and could have a very negative effect on their confidence levels.

Be patient with them, sometimes you may get incredibly frustrated because of the way they are acting, but you need to remember that they are going through something incredibly complicated and it will also be very frightening for them as they will be experiencing many emotions and feelings that they are not used to.

Even if your efforts to help are met with resistance or even rejection then you still need to keep being patient and understanding.

Try to talk to them; they may not want to tell you anything but knowing that you are there if they need to talk will be good for them, It will let them know that they do have people that care about them.

Encourage them to talk to someone else, a counsellor or a therapist could help them greatly. A study conducted by the British government showed that talking to someone that was not related to the patient in any way had a significant effect on mood levels in depressed patients.

Talking to a phone based company such as the Samaritans can also be of great assistance to them.
Understand that they may exhibit some strange behaviour whilst they are feeling depressed. According to the DSM-IV, (diagnostic and statistical manual of mental disorders) the expert's manual of mental health, some very common symptoms of depression can be very easy to brush aside. These symptoms include insomnia, forgetfulness, lack of libido, extreme happiness quickly tuning into anger or sadness and an inability to concentrate.

Remember that these are symptoms of an illness like any other and that the person suffering with them is not at fault.
But do not forget about yourself; look for support groups that can help you through this difficult time as well. Families for depression awareness and support line have people that are trained in all areas relating to depression.

How to recognise the signs of postnatal depression.

Post natal depression is something that many people know nothing about, yet according to a report by the BBC, one in ten women is diagnosed with post natal depression. But experts believe that the figure could be higher because many women do not know they have post natal depression or they are simply suffering in silence.

There are symptoms that can indicate post natal depression so if you believe you or someone else is displaying these symptoms then you or they should be checked by a medical professional. Post natal depression is not something that should be ignored; it needs to be treated.

The most common symptoms of post natal depression are:

A lack of interest in the new baby. According to a study conducted by the NHS, mothers experiencing post natal depression may have little or no interest or enthusiasm for their new baby. Some signs of this to look out for are, not wanting to talk about the baby, ignoring it when it is crying or showing no emotion when around it.

Crying is a symptom that many people brush aside because they believe that they are simply hormonal. And whilst this is understandable because there are so many hormones rushing through the body at this point, constant crying needs to be addressed. If the slightest thing starts someone crying then she may need attention.

Low moods are another common symptom of post natal depression. These moods will last a lot longer than a low mood normally would; they can last for two days to three weeks.

Irritability is also another symptom that can be easily overlooked. New mothers suffer with a severe lack of sleep and this is enough to make anybody irritable. But it could be a symptom of post natal depression if these moods last for more than a couple of days.

A lack of interest in yourself is another symptom of post natal depression, many women that suffer with this have no desire to look or feel nice. They may go for a few days without bathing or brushing their teeth.

Feeling lonely is a common symptom of post natal depression. Often the mother feels totally isolated and alone and as if she has nobody to help her. This can be hard to detect because she may not voice her concerns.

Many new mothers panic about their baby, but suffering with panic attacks is a sign of post natal depression. Signs of a panic attack are difficulty breathing, fast heartbeat, shaking, dizziness, sweating and a distinct feeling of terror.

Suicidal thoughts.

Possibly one of the biggest myths about depression is that everyone that lives with depression is suicidal and this is simply not the case. Many people that live with depression will not have suicidal thought, but some of them will.

Suicidal thoughts can be all consuming, they can leave you feeling alone, hurt, scared and helpless. And sometimes the very nature of them can mean that people don't seek help about them because these people want to die.

Suicidal thoughts can also sometimes leave you full of hatred, hatred for both yourself and other people. You might hate people for not being there for you (even though they very often don't know what you are going through) you might hate people for being happier than you, no matter what the reason, you may feel hate, and hatred is so isolating.

Suicidal thoughts can alternatively leave you numb, totally unconcerned with what happens to you. You may feel nothing, and as much as feeling nothing might not sound bad, nothing means *nothing*, no happy, no sad, no fear no excitement. Nothing.

But suicidal thoughts themselves are so dangerous, not only is there the potential for someone to act on them but the longer someone as these thoughts the harder it is to get rid of them.

Suicidal thoughts do not necessarily mean that someone is going to act on them, sometimes people only think about killing themselves. But sometimes they do it. And something that really worries me is that even to this day people *still* say that those who talk about killing themselves are the ones that don't do it. Yes they do! And even if you think they might be doing it just to get attention, is it really worth taking the risk? Of course it's not.

IF you are having suicidal thoughts, please please talk to someone about them. It doesn't have to be a professional or even someone you know (I know how difficult it is to talk to people close to you about this stuff) If you can talk to them then that's fantastic and I urge you to do it. But if you feel like you can't then talk to one of the help lines, that's what they are there for. The brilliant people on the end of those phone lines are volunteers, they are there because they WANT to help people that need to talk.

Anger.

Anger can be a big part of depression for some people, it can leave you feeling alone and misunderstood and this can leave you frustrated and frustration can very quickly turn into anger. But there are various other personal reasons why someone might feel angry.

Anger can also potentially create a vicious circle where the person feels angry and this anger shows in their behaviour, which in turn drives people away leaving the person angry again or even angrier still.

But anger can seem like a comfort blanket sometimes, because it's easier to be angry than to be depressed, at least that's what it feels like. But anger can be so hard to let go of.

So what do you do with your anger? Repress it? Talk about it? Channel it?

Repress it? No. Whenever I think of people repressing anger I always think of that episode of The Simpsons where Homer becomes the sanitation commissioner and buries the rubbish under Springfield. When the lumps come up on the golf course he tries pushing them down but they just spring up somewhere else.

And that's what happens, if you try and repress your anger about one thing, you will just start getting angry about other things. Things that might not even matter. I know from personal experience that repressing anger about something can leave you a bitter, angry and lonely person.

Talking about your anger can help, but the question remains who do you talk to? If you have an understanding person in your life then talk to them no matter who they are. Talking about your anger and hearing yourself admit that you are angry and what you are angry about can be a very liberating experience. And it can get you on the road to dealing with it constructively.

Channelling your anger can help, it can turn your destructive feelings into something positive. It doesn't matter what you do, you could sing, write, work out, anything as long as it is positive. But this is more often than not a temporary solution as what you are doing is not actually dealing with your anger, simply sublimating it and distracting yourself.

Dealing with your anger is the only way to get rid of it. Whether you do this by yourself or with the help of a professional, anger is a problem and it needs to be dealt with.

Working.

The simple fact is that some people with depression can work and some people can't. Some people's situations are just too bad for them to face the pressures of work. But for those dealing with depression and working at the same time it can be tough. Pressures, targets and expectations can potentially leave you feeling completely overwhelmed.

A good way to dealing with these feelings is to prioritise, it might sound obvious but so many people let the little insignificant things take over when they should be focusing on the things that matter.

Another thing that can potentially put people off entering the working environment is people. Being surrounded by people all day is something that can seem very daunting. Particularly as some co workers (some not all of them) can be both judgemental and gossipy.

Some good advice for people wanting to gradually get back to work (given to me by the nurse practitioner at my local surgery) is to start with some temporary voluntary work. This will not only give you the chance to interact with people in a workplace environment but it will also give you a refresher course on what it's like to be working again only without quite so much pressure.

With regards to work, those who are depressed tend to get a bad reputation thanks to the people that fake it in order to avoid work (thanks for that guys) But to the sceptics among you that say depression can't stop you working. Oh yes it can. It can leave you nervous of interacting with others, feeling as though you will not meet expectations and targets because you are not good enough and it can also leave you without any sort of motivation. There are also various other personal reasons that may leave you unable to work.

Dealing with one of those things is bad enough, but having to deal with a combination of them as many people do can be both upsetting and scary. Then there is the dilemma of if you are working, should you tell your employer or not? It should not make any sort of difference to an employer whether or not you have depression and if you feel like you are discriminated against because of it then you should consider a complaints procedure.

Also you should consider whether you want to be treated differently because of your depression, some people want their circumstances considered and want to be given certain allowances such as decreased responsibilities and the right to delegate. But conversely some people do not want to be treated any differently.

A working environment can be a minefield in itself, but when you are dealing with depression it can make the workplace that much harder to navigate.

Dating and depression.

Dating is hard enough without throwing depression into the equation. There are enough mixed emotions, raging hormones and uncertainty already even if you don't live with depression.
One of the first things people always ask me is "should I tell them?"
A valid and reasonable question but I can't tell people what to do. All I can do is try and look at it from both points of view.

Naturally the person living with depression may want their partner to know because they may have up and down days and they want them to know why. Some people also feel like their partner will think they have kept something if they don't tell them.

But even if you do not tell them then you are not keeping anything from them because that implies sinister motives. It is your situation and you tell them when you are ready and not a minute sooner.

Explaining to your partner about your situation can be scary enough as you don't know how they will react. They may freak out and run away (I can't lie it's happened to me before) or they may be completely fine with it. In fact they may well be going through a similar thing so you will never know how anyone will react to something unless you actually tell them! And if they freak out and run away then they really aren't worth running after.

But telling your partner will mean that they will understand if you seem a bit distant sometimes or you struggle with certain things. I know how scary it is but some good will always come of it.

Friends.

There are so many different kinds of friends. The main types are:
True friends
Fun only friends
Sort of friends (normally ones you have known for ages and feel you HAVE to be friends with them because you have known each other that long)
Judgemental friends
Acquaintances.

True friends are the rarest kind and if you are lucky enough to have one then never let them go. If you can talk to them about anything then do it because a true friend will help you through your worst possible times.
Fun only friends are a bit more difficult to negotiate but I had a friend who up until recently was a "fun only friend" but after a chance conversation I found out she was

dealing with depression too! So it's always worth talking to them because you never know.

We all have at least one judgemental friend, you know the one that is always right and know better than *everyone* else. Try and avoid these people as much as possible, they will only drag you down and make you feel worse about yourself.
Sort of friends are possibly the hardest ones to figure out because if you talk to them in confidence they might keep it quiet but they might not. That risk is entirely up to you.

But no matter what type of friend you are dealing with always remember that you don't have to tell anyone anything that you don't want to. It is totally your decision. And if you feel like you can't trust your friends then maybe it's time to re evaluate your Facebook list.

Families.

Families, they're complicated. And whether they are making your depression harder to deal with or you wouldn't be able to cope without them, they are there.
Talking to family members about your depression can be hard depending on how understanding or supportive they are. Some family members (I'm naming no names in mine!) feel that because they are so close to you they can say what they want to you, and whilst this is their right, it's not exactly very helpful. Hearing their judgements and criticisms is not only counter productive but it can be incredibly upsetting, especially since the whole "Idea" of family members is that they will be there for you if you need them. But sadly some of them won't, so what do we do about that?

It's not like you can avoid the ones that are always judgemental, they're your family and sometimes there's no escape! But limiting your contact with these negative people can be really helpful, even if it's only for a while. It will hopefully give you a chance to muster up some strength or to tell yourself that it's only their opinion and the only person's opinion you should be worried about is your own.

But even if your family are supportive and helpful it can still be hard to talk to them, talking to someone so close to you can feel both scary and uncomfortable. But it is always worth trying, it's always scary to tell someone how you are feeling because it leaves you feeling vulnerable and exposed. But once you have done it the feeling of relief can be so amazing, imagine releasing all of those bottled up feelings you have. Try talking to your family first, you never know they may really come through for you.

If you really feel like you can't talk to your family then there are various other people that can help. Doctors nurses and helplines are just a few places you can find the support and advice you are looking for.

Periods.

(I know any male readers will be tempted to skip over this part, but if your partner is female then this concerns you too!)

Periods are horrible, in fact depending on the person they can range from horrible to diabolical. They can be uncomfortable, painful and emotional times with hormones tearing through your body.
Periods can make you more:

Aggressive

Sensitive

Tearful

Tense

Unpredictable

Irrational

Confused

Unable to focus

It can also leave you feeling lower and more vulnerable for a few days which is totally natural but it is not what you need if you're dealing with depression.
If you find your hormones and moods are all over the place at certain times of the month then it's wise to make a note of this. Particularly if you are struggling to remain calm or positive around these times. And if the time corresponds with your period then it is worth talking to your doctor about what can be done about this. They may be able to offer you advice about how to handle your hormones.

If it is a recurring problem then please do not ignore it, periods can do really weird things to your body and although it may be natural it's certainly not very nice.

Teenagers and depression.

It seems that every time anyone mentions a teenager having depression, at least one person I know says something like "teenagers can't have depression, what have they got to be depressed about?"

And the fact is that yes they can, and not that you need a *reason* to be depressed but if you did they would have plenty! I don't care what anyone says, being a teenager is hard. Maybe it is not as hard as they make out, but it's certainly not as easy as we make it out.
The pressures on teenagers may seem minor to us, but to them they are immense. The pressure to be either what your friends and/or peers expect you to be or what your parents expect you to be can be all consuming. There is pressure to be attractive (thank you very much beauty industry) to fit in and be accepted.

And just like "kids can be so cruel" teenagers can be even crueller because they have had more time to practise. Being bullied in school can be horrific, people make light of it as adults. Imagine going into work every single day being terrified that you were going to be hurt or humiliated or mocked mercilessly. And we have the luxury of looking back on our school tormentors knowing that it wasn't the end of the world. But they do not have that luxury, in fact they do not have the luxury of hindsight for any of their problems because they are happening right now, and to them they do seem like the end of the world.

I remember being bullied in secondary school as if it was yesterday, and I hated every single second of it. I dreaded going to school because I knew I would be beaten up, humiliated, mocked and hurt. Every day, this happens every single day in school because bullies are relentless, bullies don't give you a day off.

Think back to when you were a teenager, your problems mattered to you then didn't they? So why would their problems not matter to them now?
The world can be a very lonely place for anyone, but to a teenager who doesn't fit in, is being bullied or doesn't feel good enough, the world can be just plain cruel.

But I will leave a message to those that it is happening to right now, anyone who is being bullied. It DOES get better and you WILL get through it. I know this may not seem like a great help, but one day you will look back on it and be able to say "I survived that, I can survive anything"
And there are people that you can talk to, people that know what you are going through and people that will listen and do their best to help.

Samaritans (www.samaritans.org)

National bullying helpline (http://nationalbullyinghelpline.co.uk/)

Real life accounts of what depression can do.

Some people have very kindly agreed to share their experiences of depression with me so that people can see what depression has the power to do. Names have been changed to protect the privacy of all those involved.

Jenny, 34 Manchester.

" I was living with my boyfriend of eight years and had been coping with my depression perfectly well, he was fully aware of everything that was going on with me. I was working as a project manager for a bank and was made redundant, my confidence was knocked really badly and I started to struggle. I couldn't find a job and things were getting strained financially and emotionally.

I started to sort of spiral downwards, I stopped taking care of the house and myself (I always used to take pride in my appearance but suddenly I didn't see the point any more) I had no motivation to do anything, we never went out as a couple and I very rarely went out. Things got too much for him, he could not deal with my lack of motivation or my low moods.
He even said that I was "no fun any more" I was left with a house I couldn't afford and I lost the man I had loved for the past eight years. I had to move out of my beautiful home and into a tiny little flat and have it paid for by housing benefits, I was on E.S.A (employment support allowance) I hated the fact that I was relying on someone else to give me money but I really couldn't face working. I felt totally useless and worthless. I had my dosage of anti depressants upped from 30 mg to 90mg. I was tired all the time and I felt like I was numb.

Elaine 29 from Milton Keynes.

I moved out of my Mum's house when I was 23 to start a new life with my boyfriend. I moved over 300 miles leaving everyone I knew behind. I had been on anti depressants for a few years but never had any what I call "major" problems. Things very quickly changed with my boyfriend as soon as I moved in. What had once been a fun loving man turned into a raging control freak. Nothing was ever good enough, I had a job and earned money, but it was not enough, I cleaned the house, but I never did it right. I missed my family and friends, but he was always "too busy" to come with me to see them. I visited them on my own and made excuses each time for why he wasn't with me. After I found out he was cheating on me I told him I was leaving, he begged me to stay and told me that he would finish it with this other girl and that he loved me.

Like a moron I agreed to stay. And things changed for a while but he quickly became secretive and distant. I knew he was cheating again and eventually he even stopped trying to hide it. I knew virtually no one where I was now and I felt so alone and so hurt. All I could think of all day was ‚Why didn't he love me?, what was so wrong with me? What is so great about her? These feelings became my entire life. I was so down I could hardly be bothered with anything, I had no confidence left and I had the bare minimum contact with my friends and family.

One day whilst he was texting her, I made a decision. I was leaving. The next day whilst he was at work I packed as many of my things as I could fit in a bag and got on a train back to Milton Keynes. I was so down about the whole thing for months but with the help of my friends and family I got a new job, my own flat and even started dating again a lot sooner than I thought I would.

Louise 25 from Glasgow.

I had been depressed since I was about fourteen, I never told anyone at all until I was eighteen and I tried to kill myself. I took an overdose and cut my wrists at the same time, neither of them worked. I was in hospital for four days and they made me see the head psychiatrist. I didn't want to talk to her so I just said I didn't want to kill myself any more, even though I kind of did. I went on with my life, still not having spoken to anyone and about a year later I tried again. I tried to kill myself five times in total over a period of four years. Two times nobody knew about, I tried to hang myself once and the bar I tied the rope to fell off the wall, the second time I took an iron overdose and seemed to have no bad effects at all.
People might think it's stupid but I felt like after all those failed attempts I was still here so there must be a reason. I had a purpose so I just had to figure out what it was.

I spoke to my doctor one day about it, it was so much easier than I thought, once I started talking I couldn't seem to stop. We decided on a course of treatment for me, together, she was really understanding and sympathetic and took my feelings into consideration. I started counselling and my counsellor Dave was a great guy, I was sceptical about having a male counsellor at first but he was brilliant.

I started working and volunteering for MIND at the weekends. I got my own flat, a great job and am working towards my own counselling qualification. I have a fantastic girlfriend who understands about everything I went through, it's been a slow process but looking back on the person I used to be is like looking at someone else's life!

Kate, Louise's Mum.

It nearly killed me when Louise tried to take her own life. She must have been so frightened and in so much pain to want to do something like that. I thought she could talk to me about anything and I felt like I had failed her. I wanted her to talk to me after what happened but every time I tried she just clammed up. It scared the hell out of me, every night when I went to sleep all I could think was that I was going to go upstairs in the morning and find her dead. That was such a terrifying thing to think that every time I said goodnight to her, it might be the last time I saw her again, my child.

When it kept happening I wanted so much to send her to hospital for a while to get her the help she needed, but I knew if I did that I may never see her again.

I remember her saying one day that she was going to the doctor, and when she came back, she looked different. She actually looked different. And from then onwards it was all uphill, she had some relapses, but she dealt with them so brilliantly. I have never asked her why she tried to kill herself, she will tell me if and when she is ready. But I am so proud of her for getting the help and coming out the other side.

Marie, 61 Watford.

My husband was one of the happiest, funniest men alive. Right up until he went into our garage, locked the door and killed himself in the car. He left me with five kids to raise (It was 26 years ago) and a husband to mourn. I had NO idea he was having any thoughts like this, he never spoke to me about it once. He just did it. Just killed himself. Everyone said he was the life and soul and he always had a joke ready, but it was all an act. I feel like the husband I knew was a lie. I didn't know him at all. And as much as I loved him and still love him, I will never be able to forgive him.

Kevin, 40, Oxfordshire.

For years I knew I had depression, but I never told a soul. If it had got out I felt like I would have been a laughing stick, I mean, men don't get depressed do they? But they do, I was so down all the time, I was so unhappy but I never let it show. I was always "happy" and "up for a laugh" but I wasn't I was sad and desperately lonely. I had a brilliant wife who looked after our amazing kids, but I couldn't tell her about it. Something had to give though, I was struggling at work, losing my friends and I was mentally exhausted ALL the time.

In a desperate attempt to try and conquer this,I called the Samaritans. I was expecting them to tell me I was tying up the phone lines and I wasn't an emergency, I wasn't suicidal I was depressed. It was the best decision I ever made, the man I spoke to was helpful, friendly and he just listened. He listened to what I had to say, he didn't tell me what to do, he just listened. After I hung up I could already feel myself becoming a tiny bit more positive, I had unburdened myself and it felt like such a relief! After a few more calls to them I decided that I had to talk to my wife, she deserved to know what was going on. She was fantastic about it and even had a go at me for not telling her sooner! Now I talk to her and the Samaritans and am well on the way to being my old self again.

Everyone's experience of depression is different, some people can go on with their lives, but some have their lives devastated by it. There is no "textbook case of depression"

Getting back to "normal"

When you live with depression, sometimes it's hard to imagine that you will ever have a normal life again. That you will enjoy the things you once did and be the person you once were. But it can happen, it will not happen quickly and it will not be easy, but with the right help and the right outlook it will happen.

Yes you will have your down days, everyone does and there will be times when you are getting better when you wonder if it's all worth it and if there is any point in carrying on. But there is. You want to get through it and you want to survive, you got help so you want to get better.

But your life WILL be your own again and you will not feel like your life and everyone else's is ruled by your depression. I am not saying that you will forget that you ever had it, because you won't and you shouldn't. It can teach you some pretty valuable lessons which are not always clear at the time but they will come to light eventually. Don't dwell on your depression and don't let it define you, but don't pretend it never happened. Remembering what it can do to you is one of the best ways to help stop it happening again.

My life was totally destroyed by my depression; And it's been the most difficult task imaginable trying to get my life back to something like normal. And when I am struggling I force myself to think back to everything terrible that happened, and it's my biggest motivator because I never want to be that low again.

I know what it's like to feel like there is no point in you being alive. But everyone has a purpose. Everyone has something to live for, something to reach for and something to care for.

Thank you.

I talked briefly about medical professionals and their opinions and various treatments of depression, but I just wanted to say how lucky I have been with the people that helped me through mine. I received excellent, compassionate advice and support from almost everyone involved.

But I would particularly like to thank one of the nurse practitioners from my local G.P's surgery, obviously I can't name her but throughout my treatment she has been kind, concerned, professional, thoughtful and helpful and I can honestly say I don't know what I would have done without her.

I would also like to thank my wonderful partner Claire Missen for her support and faith (and for letting me write at hers when my neighbours were being too loud!) Sarah "happy" Gilmour for answering my random texts about titles and Kerry Millward for just being there with a drink and some good advice.

www.ingramcontent.com/pod-product-compliance
Lightning Source LLC
Chambersburg PA
CBHW021856170526
45157CB00006B/2473